What Can You Do, Kangaroo?

Written by Jenny Phillips

Illustrated by Ecaterina Leascenco

Cover design by Phillip Colhouer

© 2020 Jenny Phillips

goodandbeautiful.com

Hello, kangaroo! What can you do?

I can carry my baby in a pouch.

I can hop and jump! This is how high a boy can jump, and this is how high I can jump.

Hello, bear! What are you doing in there?

I am sleeping in a cave all winter with my cubs. I will wake up in spring.

Then my cubs and I will eat berries and grass.

I will also catch and eat fish.

Hello, fox in the snow!
How fast can you go?

I can run as fast as a car.
I am speedy as can be!

I sleep in my den in the ground.

I am usually awake at night.

Hello, elephant so wide and tall!
Can you make a really loud call?

Yes, I make calls with my long, long trunk. I use my trunk to suck up water to drink. My trunk can also smell and hold things.

I live in a family called a herd.

I love to swim and bathe
and play in water.

**Hello, monkeys swinging around!
Do you ever walk on the ground?**

**Most of us monkeys spend
our time in the trees, but we
can all walk on the ground.**

We eat seeds, fruits, flowers, nuts, and insects.

There are many, many types of monkeys. Here are some of them.

Hello, camel in the golden sand. How do you like life on the desert land?

God made my body perfect for places that are hot and dry.

I love to be useful. I can carry heavy loads.

Hello, lion now fast asleep. Can you run and can you leap?

Yes, I like to sleep in the shade or even in a tree, but when I'm awake, I can run very quickly and leap very far.

My roar is so loud it can be heard from far away.

I need to eat a lot of meat, so I go hunting all the time.

Hello, tiger with stripes of black.
May I ride upon your back?

No, no, no. You would not want
to ride on me. I am the biggest
wild cat in the world, and it
is not safe to be near me!

I am kind to my babies. I have two or more babies at a time. They are my cubs.

Hello, hippo! What kind of things do you know?

I know how to hold my breath for up to seven minutes.

I spend most of my time in the water, but I eat grass when I go on land. Don't get close to me! I am the most dangerous land mammal in the world.

Hello, giraffe with your head in the tree. Are you actually eating those leaves I see?

Oh, yes, I love to eat leaves. I also love to eat twigs, do you?

I am very tall, and my neck is very long.

I also have a dark blue tongue that is very long.

Hello, little squirrel! Can you run and jump and twirl?

Yes, I can jump over 6 feet high, which is taller than you! I can jump from tree to tree, or even straight up in the air.

I live high in trees to keep me safe, and you can find me almost everywhere in the world.

When my babies are born, they cannot see. I keep them safe while they grow.

Hello, panda, I see you chew.
Will you eat all that bamboo?

I eat this bamboo all day long; I need lots and lots to keep me going—sometimes 40 pounds or more!

I like to live all by myself, and you can only find me in the mountains of China.

Hello, deer, why are you here?

We are here to eat the soft green grass, but we are always ready to run away fast.

My baby is called a fawn. He has white spots all over his body to protect him and make him harder to see.

This deer has antlers. Each year his antlers will fall off, and he will grow a new, bigger pair.